The Jarrold Nature Series

Common British Lichens

Text and photographs by **Frank S. Dobson**

Jarrold Colour Publications, Norwich

Lichens are a group of organisms that are not really comparable to flowering plants, ferns or mosses. They are fungi which obtain the sugars necessary for their existence not by living on decaying materials or attacking live organisms but by forming special stable associations with algae. In conjunction with the algae the lichen-forming fungi produce a vast array of variously coloured crusted, tufted, leafy and pendent structures designed to display the enclosed algae to the maximum benefit of the fungus. Lichens are extremely slow growing, most increasing only a few millimetres each year, but at the same time they are exceptionally long lived. Some may reach several thousand years old making the collective common name (time stain) particularly apt. Precise environmental conditions are needed to maintain the fungus-alga association without disturbing the equilibrium and they are able to thrive in many habitats few other organisms can colonise, such as on trees, fences, rocks and sand. Some species even grow several millimetres into rock, often with only their fruiting bodies visible on the surface. Lichens are found throughout the world and in regions with harsh climatic conditions they often greatly outnumber the flowering plants. Some lichens are capable of being dried for many months and then, on being wetted, at once start to photosynthesise as vigorously as before.

The fruiting bodies (ascocarps) of lichens are mainly disc-shaped (apothecia), in lines (lirellae), or flask-shaped (perithecia). If a section of a fruit is examined under a microscope, the spores will be seen contained in a transparent sac (ascus). These spores vary in shape and size but are all very small, usually less than one thousandth of an inch long. Unlike most other fungi, each fruiting body may continue to liberate spores for many years. These spores come only from the fungal partner, and if they are going to grow into a lichen, they must after germination, rapidly capture some cells of the appropriate alga or die.

Many lichens are dispersed by pieces breaking off and forming new plants. In addition, finger-like extensions (isidia) or powdery granules (soredia) which contain both algal and fungal cells are often formed on the lichen. These are the most important means of dispersal in many lichens. They are dispersed by the wind and rain and also by insects, a number of which eat lichens and then carry off fragments adhering to their bodies. If these fragments land on a suitable site they may grow into a new plant. Birds carry off the more shrubby lichens to line their nests and so may transport species considerable distances. Despite this, if an area is clear-felled it may prove impossible for some species to return to a site due to the distance involved. It is vital in sites containing important lichens, that some mature trees bearing the lichens are left standing to enable younger trees to be recolonised.

Man has had many uses for lichens, and they had an important place in many herbal remedies. *The Doctrine of Signatures* was an important concept in these herbals. Under this system, it was believed that if a plant resembled the shape of a part of the body, or even had the same colour or smell, this was an indication that it should be used to cure diseases of that organ. It has recently been found that a number of them do in fact have antibiotic properties and some have been marketed in Scandinavia and the U.S.S.R., but the slow regeneration of lichens prevents any very large scale industry. They were once a major source of dyes and are still used for this purpose and also as a fixative in the manufacture of perfume. Lichens provide the largest constituent of the winter food of reindeer and caribou. Sir John Franklin considered that the lives of the members of his expedition were saved by eating lichens, and there have been a number of more recent instances of shipwrecks and plane crashes, where lichens have assisted in providing food for survival. Among exotic uses they were brewed into beer, formed the stuffing in mummies and were even used as a smoking mixture.

Many lichens can be recognised with the unaided eye but with a lens many interesting details that would otherwise be missed are apparent. The following descriptions cover only a small proportion of the nearly 1,400 species of British lichens and are divided into coastal and inland species – but a number of the species may be found in either situation.

There are many books available dealing with lichens, but the following will be of particular interest to the novice:

Alvin, K. L. (1977) **The Observer's Book of Lichens.** *Warne.* Descriptions, keys, monochrome and coloured illustrations to the commonest species.

Duncan, U. K. (1970) **Introduction to British Lichens.** *The Richmond Publishing Co. Ltd.* Descriptions and keys to the common and many rare British lichens.

Dobson, F. S. (1979) **Lichens. An Illustrated Guide.** *The Richmond Publishing Co. Ltd.* Illustrated keys, descriptions, distribution maps and photographs for the commoner as well as some rare British lichens.

Hawksworth, D. L. and Rose, F. (1976) **Lichens as Pollution Monitors.** *Arnold.* Includes notes on methods of study of lichens especially in relation to air pollution.

Richardson, D. H. S. (1975) **The Vanishing Lichens.** *David & Charles.* An account of their history, biology, uses and conservation.

If you wish to learn more about lichens, both novices and advanced students are welcome to become members of The British Lichen Society, c/o Dept. of Botany, British Museum (Natural History), Cromwell Road, London SW7 5BD.

The scale of reproduction of the photographs has been printed next to each illustration except where it is obvious or approximately life size.

Lichens of the Seashore

Lichens often form a series of clearly defined communities that start from well below the high tide line where they are covered by the sea for much of the time, through the region of heavy wave action and spray, until they merge with inland species. It is usually possible to see three distinctly coloured zones of lichens on a rocky seashore. The lowest is a black crust-like layer which reaches from below the high tide mark up into the region where it is still subjected to heavy spray. In this zone only the crust-like lichens are able to survive and not be washed off the rocks. Above this in the region of only light spray is an orange layer where the first of the leaf-like lichens are able to survive. This orange zone frequently merges into a grey zone situated above the majority of the spray. Although the grey effect is due mainly to grey crust-like lichens, it is in this zone that the larger leaf-like and shrubby lichens are able to grow without being washed away.

1

1. Lichen zones on a rocky sea coast. The illustration clearly shows the black, orange and grey lichen zones. The width of these bands varies with the exposure to the weather of the seashore. The zones tend to be wider and start higher up the rocks as the shore becomes less sheltered or when it is south-facing and therefore dries out more rapidly. Cliffs of soft chalk may erode so rapidly that lichens are prevented from establishing themselves in this habitat.

2. Tar Lichen *(Verrucaria maura)*. This lichen usually forms the main constituent of at least the upper part of the black zone. Growing from well below the high tide mark it may cover the rocks with a black gelatinous crust that looks like dried tar. If examined with a lens, the small fruits may be seen scattered over the cracked surface of the lichen.

3. Orange Sea Lichen *(Caloplaca marina)*. One of the first orange lichens to be found above the high tide mark. It is a deep yellow-orange and usually consists of numerous small granular pieces with deep reddish-orange convex fruits with paler margins scattered among them. It may become well developed with small convex lobes that form rosettes.

2

3

3 × life size

4. Orange Sea Star *(Caloplaca thallincola).* This lichen grows on seashore rocks above the tide line often being found together with the Orange Sea Lichen. It can be easily distinguished from that lichen as it forms distinct neat rosettes with long convex lobes that grow out almost from the centre of the plant and it is usually a lighter orange. The fruits are slightly convex, orange with a paler margin. It is common on rocky shores throughout Britain.

5. Common Orange Lichen *(Xanthoria parietina).* Orange-yellow becoming green-yellow in the shade, the centre usually has numerous bright orange fruits. This is a very common lichen on rocks, walls, sea defences, etc. and it is also common inland, even in areas of moderate air pollution. It prefers nutrient-enriched sites and the Eskimoes use this fact when they are hunting marmots, looking for the yellow splashes of the lichen on the fouled rock under marmot holes.

6. Warted Orange Lichen *(Xanthoria aureola).* Very similar to the previous species, except that it is often deeper orange and the lobes in the centre of the plant are covered with fine warts which give it a granular appearance. It is common in the east of Britain, becoming much less so in the west. Due to its yellow colour, it was at one time taken as a supposed cure for jaundice. Like many other yellow lichens it was used to produce a yellow or brown dye.

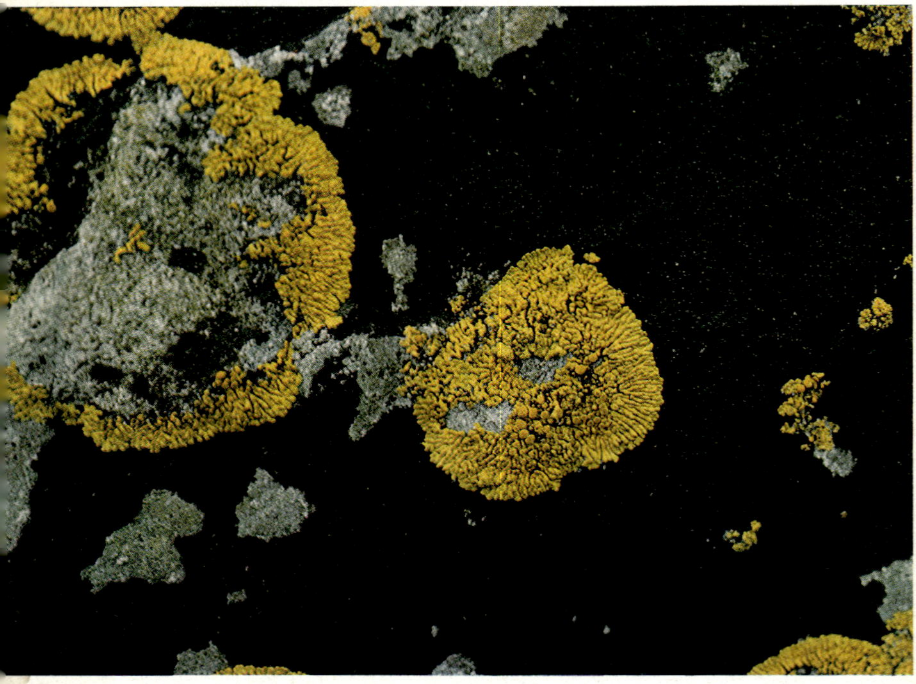

4

5
6

7 ½ × *life size*

8

7. Sea Ivory *(Ramalina siliquosa).* The shiny yellow-grey to greenish-grey tufts of scimitar-shaped lobes of this common lichen are found growing on hard seashore rocks where it is out of the reach of the waves. It is variable in its appearance, and is sometimes very contorted and warty. It is always brittle and is very difficult to remove from the rocks in one piece. It often has pale fawn coloured fruits growing on the edges of the lobes or towards the tips.

8. Narrow Lobed Rock Lichen *(Anaptychia fusca).* The narrow, matt brown, radiating lobes of this plant form large rosettes on marine rocks, the lobes often bearing dark brown disc-shaped fruits. It is one of the first of the leafy lichens to be found above high water mark, and is found right around Britain, only being absent where there is a lack of suitable rock. In the west it also sometimes grows inland on stone walls and trees.

9. Black Shield Lichen *(Lecanora atra).* Frequently found forming grey or white crusts on hard siliceous rocks, this is a common lichen of the grey zone. The thick cracked crust is often rough and warted. The black discs of the fruits have thick grey rims and the fruits are often grouped towards the centre of the plant, where they contrast strongly with the light grey of the crust. It is also common inland on rocks and walls.

9

2× life size

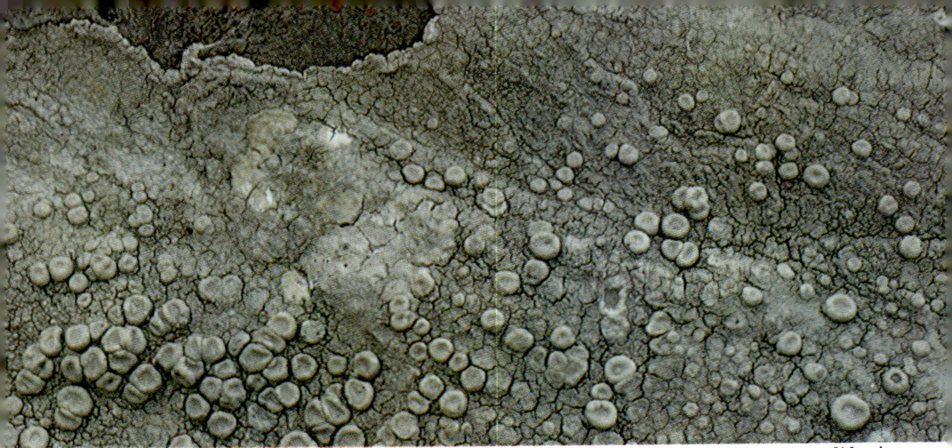

10 2× *life size*

11

½× *life size*

10. Parelle *(Ochrolechia parella)*. This species forms a grey or buff crust on coastal or inland rocks. A dye was produced from it by the action of ammonia, for which the cheapest source was stale urine. This used to be collected and sold, but beer drinkers were discouraged as they were said to produce too much water and not enough ammonia. It has been said that this was the cause of the distinctive smell of old lichen-dyed tweed.

11. Orchil *(Roccella fuciformis)*. On sheltered overhanging rocks near the sea in the south-west may be found the pale lavender-grey fronds of this rare (not to be collected) lichen. The secret of making a dye from similar species of lichens was lost, and then rediscovered in the fourteenth century by a Florentine merchant, who found by accident that the action of stale urine produced a purple dye. Litmus is still produced from a closely related subtropical species.

Inland Lichens

As on the coast, inland lichen communities depend on climatic factors, degree of pollution, etc. Even the amount of light affects the species present, some being found only in shaded woodlands while others may only occur on exposed heaths. There must also be a suitable substrate available, for although in suitable conditions many species will grow on a wide range of trees or rocks, others must have the correct acid or calcareous rocks, or type of tree before they can develop. With a little experience it is soon possible to estimate which lichens are most likely to be found at any particular site.

This section has been subdivided into the species found mainly on trees or rocks, in woods, hedges or parkland. This is followed by those species that are frequently found in areas of high air pollution. The final section covers the lichens of moorland or coastal dunes. Many species are common to both those regions of rather poor soil.

12

½ × life size

12. Tree Lungwort *(Lobaria pulmonaria)*. Bright green when wet but becoming khaki-coloured when dry. The lobes have raised, net-like ribs, which often have fine granular outgrowths (isidia or soredia) along them. The lower surface of the lobes is light brown and felt-like. This lichen is now only found in areas of very clean air which have a continuity of ancient trees. Due to its lung-like shape it was once used as a supposed cure for tuberculosis.

13. The Script Lichens *(Graphis scripta* and *G. elegans).* On the smooth bark of shaded trees, the white to grey patches of these lichens can be easily noticed. These patches have on them dark lines resembling a strange form of writing. These are the fruits, which in this species are not wide discs but slits with hard carbonaceous margins. When growing on young trees the growth of the tree often pulls these fruits out into parallel lines.

14. Mustard Powder Lichen *(Lepraria candelaris).* The bright mustard-yellow colour of this lichen looks at a distance like streaks of yellow paint. On close inspection it will be discovered that this lichen consists of a structureless powdery layer in the dry cracks of rough-barked trees. It has never been found in fruit and dispersal must be by the small powdery pieces breaking off and being carried away by the wind, insects, etc. to a new host.

15. Netted Shield Lichen *(Parmelia sulcata).* A common plant on trees, especially in lowland Britain. It is a large lobed lichen, pale grey, or greenish grey when in the shade. The lobes are slightly wrinkled and widen out at the tips. Under a lens a network of white spots and lines may be seen on the lobes. These may develop a powdery growth (soredia). It is usually found on trees to which it is attached by black root-like rhizinae.

3× life size

14

2× life size 15

16
17

16. Fringed Leaf Lichen *(Parmelia perlata).* A fairly common lichen on trees and sometimes rocks in unpolluted areas. It has large lobes but, unlike the previous species, the powdery soredia arise only around the edges of the rather erect smooth lobes. There are often stout black hairs growing out from the edges of the lobes. Like many other *Parmelia* species it was often used to produce a brown dye.

17. Puffed Shield Lichen *(Hypogymnia physodes).* This lichen appears to be able to withstand more air pollution than most leafy lichens. Unlike other leafy lichens, the lobes are hollow. In mature plants the ends of the lobes split open and turn back exposing the powdery soredia. It is attached directly to the surface on which it is growing and lacks the root-like rhizines of many other species.

18. The Wrinkled Shield Lichen *(Parmelia caperata)*. When wet it is a bright, almost apple-green becoming yellow-green when dry. This is a common lichen but it is not found in the north. It derives its name from the transverse wrinkles that occur towards the centre of the plant. The lobes have yellow-green granular soredia which may cover the whole plant.

19. Yellow Boulder Lichen *(Parmelia conspersa)*. This is found mainly on acid rocks in upland areas, but it is also found down to sea level, and is shown here growing with other upland species. Like the previous species it is yellow-green in colour but it differs in having fine rod-like isidia instead of the granular soredia. The undersurface is black with root-like rhizinae which become white at the paler lobe tips.

18

$1\frac{1}{2}\times$ life size

19

20

21

20. The Dog Lichens *(Peltigera canina* and *Peltigera praetextata).* Large leafy lichens up to 6 in. or more across. Brown when wet, grey when dry, often bearing red-brown fruits. They are common in woods, sand dunes and open spaces and are often found on damp mossy banks. The pale felted lower surface has spike-like projections that resemble dogs' teeth and from these it gets its name and was at one time tried as a cure for rabies.

21. Glaucous Leafy Lichen *(Platismatia glauca).* Able to withstand slight to moderate pollution, this is a common lichen throughout Britain, growing in acid situations on trees and rocks. It has erect thin and papery lobes which are greenish grey, often with faint white dots and lines. The undersurface has no or very few 'root hairs' and is light brown or almost white. The edges of the lobes become covered in rod-like isidia or powdery soredia.

22. Old Man's Beard *(Usnea florida).* A common sight on trees in unpolluted areas, this lichen is usually found with numerous pale flesh-coloured fruits, fringed with long branchlets. It formed the basis of the famous 'cyprus powder', used during the seventeenth century as a perfumed hair powder. In this photograph it is surrounded by the Lesser Old Man's Beard lichen (*Usnea subflorinada*); this is a common but variable lichen which is rather rarely found in fruit.

23. Oak Moss *(Evernia prunastri).* This yellow to green-grey strap-shaped lichen has a white cottony undersurface and is very common except in northern Scotland. It is mainly found on deciduous trees; on the trunk or, as shown here, even on the fine twigs. This species has had many uses such as a primitive antibiotic or to make a hair dye, and it is still used as a fixative in perfume. It is also favoured by long-tailed tits for nest-building.

24. Ragged Mealy Lichen *(Ramalina farinacea).* A common species that is found on nutrient-rich bark throughout Britain. It differs from Oak Moss in being greenish grey on both the upper and lower surfaces. The edges of the shiny strap-shaped branches split open and in these splits powdery soredia are formed. *Ramalina* species were at one time ground down to produce a white hair powder and it was also tried as a cure for chilblains.

25. Golden Hair Lichen *(Teloschistes flavicans).* Britain is at the very north of the range of this pollution-sensitive lichen and it is now found only in the south-west and Wales. Its yellow-green to orange fronds make it clearly visible on the nutrient-rich trees and rocks that it prefers. It is found mainly in maritime areas and on rocks by the seashore. It has been used to produce a yellow dye but due to its rarity it should not be collected.

2× life size

24

25

Lichens and Air Pollution

Lichen species are sensitive to various pollutants of which sulphur dioxide air pollution is the most widespread. Using the different tolerances of various species it is possible to devise scales to estimate the mean sulphur dioxide levels in the air. An example of such a scale is given below. This scale uses species that grow on trees, but others are available for rocks and walls in urban areas. For details of such scales see Hawksworth and Rose (1976), *Lichens as pollution monitors.* It must, however, be remembered that the absence of a species can be caused by many other factors, such as climate, other pollutants or the lack of a suitable substrate. The centres of dense woods and the floors of narrow valleys will often be found to have significantly lower levels of pollution. Some lichens in levels B and C have managed to take over the ecological niche left vacant when the more sensitive leafy and bushy species were killed off by increasing pollution.

(Based on the zone scale of Hawksworth and Rose)

Level	Species on Trees	Mean Winter SO_2 ($\mu g/m^3$)
A	No lichens	Above about 170
B	Lecanora conizaeoides (B)	150
C	Lecanora conizaeoides, Lepraria incana, Buellia canescens (N)	125
D	Hypogymnia physodes (B), Parmelia saxatilis (B), P. sulcata (B), Xanthoria parietina (N)	70
E	Hypogymnia physodes, Parmelia saxatilis, P. sulcata, Lepraria candelaris, Evernia prunastri (B), Plastismatia glauca	60
F	Parmelia caperata (B), Graphis scripta	50
G	Parmelia caperata, Usnea subfloridana	40
H	Lobaria pulmonaria, Usnea florida, Parmelia perlata, Teloschistes flavicans	less than 35

(B) *Found only on the lowest 2–3 feet of the tree*
(N) *Found in nutrient-enriched sites*

Old churchyards are very important sites for lichens, the gravestones are often made from a variety of rocks, many of which may be the only suitable site in the locality for a particular species. This valuable habitat is unfortunately frequently being destroyed by removing or cleaning gravestones and placing them around the edge of the churchyard. When this is done, the alteration in micro-climate, light and other factors will, in many cases, cause the lichens to die off, followed by the loss of associated invertebrates such as mites, moths, etc.

26. Grey Crust Lichen *(Lecanora conizaeoides).* Until early this century this species was rarely found, and it is still rare in unpolluted regions. It has a low competitive ability but due to its pollution resistance it has taken over the trees which have lost their normal lichen cover due to pollution. Its grey-green granular crust covers the lower parts of many trees in lowland Britain, even in urban areas.

27. Green Powder Lichen *(Lepraria incana).* A light green powdery lichen without any definite shape, growing on shaded rocks and trees throughout Britain. It is one of the most pollution-resistant species and even manages to grow on damp tree trunks in city parks. This resistance to pollution is probably due, at least in part, to its ability to remain unwetted. This will be seen in the photograph in the drop of water standing on its surface.

26 *6 × life size*

27

28. The Bordered Wall Lichen *(Lecanora muralis).* This is a fairly rare lichen in unpolluted regions. However, it is now very common in many polluted areas, especially on man-made substances such as asbestos-cement and tiles. These are alkaline and this helps to neutralise the air pollution. The brownish-green colour of the lichen is distinctive, and the overlapping small, white-margined lobes and its many fruits aid recognition.

29. Grey Warted Lichen *(Physcia caesia).* This lichen is fairly resistant to pollution, and forms light grey patches on asbestos roofs, concrete, tombstones, etc. It likes fairly well-lit sites that are nutrient-rich through bird droppings or fertiliser. It has long, narrow lobes and scattered over the plant are globular powdery growths (soredia); these are yellowish grey to blue, often with darker blue flecks.

30. The Mustard Lichen *(Candelariella vitellina).* Although usually small and often overlooked, the golden to mustard-yellow granules and fruits of this lichen may even be found in areas of fairly heavy air pollution. The small granules of which it consists sometimes clump together, and form a crust on nutrient-rich rocks and walls. In Sweden a yellow dye was extracted from this lichen, and used to dye candles for use in church services.

28

2× life size

3 × life size **29**
4 × life size **30**

31 *2× life size*

32 *2× life size*

31. The Frosted Wall Lichen *(Buellia canescens).* This is very common in the south, becoming rare in north Scotland. It can withstand moderate air pollution and is found on nutrient-rich trees, walls and rocks. Its pale grey lobes have a frosted appearance (pruina), and the centre of the plant is usually covered in a floury mass of soredia. Except in the extreme south it is seldom found bearing its black centred disc-shaped fruits.

32. Orange Star Lichen *(Caloplaca heppiana).* The bright orange rosettes of this lichen are common on hard limestone and especially tombstones; more rarely it also grows on brick. It is found throughout lowland Britain but becomes local in the north and west. The numerous fruits in the centre of the plant have orange discs with a paler margin. Towards the tips of the lobes, there is a heavy dusting of a white powder (pruina).

33. Sulphur Lichen *(Lecidea lucida).* The bright yellow-green patches of this species belong to one of the few species that can be easily recognised at a distance, but at close range it will be seen to be a structureless mass. It is common throughout the country, often following the lines of mortar or individual bricks or, as shown in the photograph, it will sometimes pick out an inscription on a tombstone by growing in the slightly damper carving.

33

Lichens of Dunes and Upland moors

The difficult conditions found in these regions suit many lichens. Both dunes and moorland usually have a poor soil and frequently either too much or too little water for the growth of most plants. Lichens are, therefore, able to escape from much of the competition from the more vigorous flowering plants. The poor soil also discourages man from destroying the habitat by ploughing it up to plant crops, and the thin grass can support only a few animals. Even the exposed rocks in unpolluted areas are often completely covered in a mosaic of brightly coloured lichens. The soil and rocks on many moors can be very acid and this suits a number of species. Sand dunes can also be rather acid but some, enriched by sea shells, provide a chalky substrate that is colonised by different lichen species from the acid sand.

As many species of lichen grow in a variety of habitats, other sections of the book should also be consulted to identify a specimen.

34

2× life size

34. Common Crottle *(Parmelia saxatilis).* A very common lichen of mainly upland regions, found loosely attached to rocks and trees. The pale grey to bluish-grey lobes resemble the Netted Shield Lichen in the white spots and lines on the lobes. It differs in the fine rod-like outgrowths (isidia) growing on these lines. At one time, if it could be gathered from the skull of an executed criminal it fetched its weight in gold as a magical drug.

2× life size 35

36

35. Black Crottle *(Parmelia omphalodes)*. Very common on acid rocks in upland areas but it extends down to sea level. It is brown to grey-brown, often having a rather metallic purple tinge. This lichen has pale spots and lines on the lobes but it lacks soredia or isidia, and is more commonly found with disc-shaped fruits than either Common Crottle or the Netted Shield Lichen. It was used to produce a brown dye which was said also to mothproof the garment.

36. Devil's Matches *(Cladonia floerkeana)*. A close search of sand dunes or more especially peat moors will often reveal this lichen where it provides a splash of colour on the brown peat of an upland moor. It has a blood-red fruiting body capping a light to dark ash-grey (greenish when wet) stalk. This stalk may be branched towards the tip but is more commonly unbranched and from the resemblance to a red-headed match it gets its common name.

37

38

37. Cock's Comb Lichen *(Cladonia coccifera).* Found on dunes and moors, often with the previous species, and is shown here in a typical moorland lichen community, but it also grows on soil pockets in walls, mainly in upland regions. It differs from Devil's Matches in the yellow-green, rather granular stalk which gradually widens into an irregular cup. The bright red fruits form around the rim but in mature plants may cover the whole of the top of the cup.

38. Slender Pixie-cups *(Cladonia fimbriata).* There are a number of pixie-cup lichens; these vary in shape and roughness of the cup. The one shown in the photograph is very elegant with a 'golf tee' shape, the long stalk opening out abruptly into the cup. It is very common on sand dunes as well as on soil and tree roots in hilly regions. Other species of pixie-cup are found throughout Britain wherever the air is not too polluted.

39. Reindeer Moss *(Cladonia impexa).* This is one of a group of lichens called Reindeer Moss. This name is, however, mainly applied to the rarer *Cladonia rangiferina.* In Arctic and other cold regions they do indeed form an important source of food for reindeer, caribou, etc. but due to their slow growth, it may be necessary to leave as long as ten years between each grazing. They are often dyed and sold as trees for model railways.

40. Elk's Horn Lichen *(Cladonia foliacea).* Common mainly in short turf in coastal areas, especially on cliff-tops and sand dunes, or inland, particularly in Breckland. The small green leaf-like lobes are often clustered together in patches several inches across or they may be found as isolated lobes. In dry weather, they become more conspicuous as the lobes dry out and curl over and expose the pale yellowish-white undersurface.

39

40

41. Pink Peat-Lichen *(Icmadophila ericetorum).* This is one of the first plants to colonise fresh saturated peat where it forms patches of greenish-grey to white granules. The pale pink fruits develop quickly and become convex and bun-shaped. It is common in upland regions especially in the north and in Scotland. It was formerly widespread, but is now extinct throughout much of lowland Britain due to the drainage of bogs and heaths.

42. Red Eye Lichen *(Haematomma ventosum).* This lichen is found growing on exposed, well-lit rock faces or boulders in mountainous areas. It has a solid, smooth or warted crust which may be either grey or more commonly yellow-green, and often outlined by a thin dark margin. It obtains its name from the large fruits which are the colour of dried blood and are usually surrounded by a pale margin. It was at one time scrapped off the rocks to produce a rust-brown dye.

43. The Map Lichen *(Rhizocarpon geographicum).* The bright yellow to greenish-yellow colour of this very common lichen is very distinctive and at close range the black lines and fruits that give it its name can be seen. It grows on hard acid rocks mainly in upland and coastal regions. Recently it was found in Norfolk far from its nearest known site still growing on a slate on a church roof, very many years after that had been brought from the original quarry.

41

2× life size

3 × *life size*

42

43

44
45

44. Fringed Rock Tripe *(Umbilicaria cylindrica)*. There are a number of species of rock tripe found in upland Britain. They are all rather flat and leathery and are found on bare rocks attached only at one central point. This species is battleship-grey to brown and the lobes have a fringe of rigid black hair. The centre often has a pale frosted look. Fruits are common and under a lens they look like small black Chelsea buns.

45. Fleecy Rock Tripe *(Umbilicaria polyrrhiza)*. This usually consists of only a few rather thick, dark brown (green when wet) shiny lobes. The undersurface is dense black and covered with numerous root-like rhizinae. The Many Lobed Rock Tripe (*Umbilicaria polyphylla*) is also common but lacks the root hairs and, as its name suggests, it usually has many small lobes. Rock tripes have been boiled and eaten as a survival food but they provide little nourishment.